COTTON TO CLOTHING

Robin Johnson

A Crabtree Seedlings Book

Crabtree Publishing
crabtreebooks.com

Author: Robin Johnson

Editor: Ellen Rodger

Proofreader: Melissa Boyce

Design: Katherine Berti

Photo research: Robin Johnson, Katherine Berti

Prepress and print coordinator: Katherine Berti

Photographs and illustrations:

Istockphoto
- dszc: p. 10 (bottom)
- wsfurlan: p. 13 (bottom)

Shutterstock
- Akella Srinivas Ramalingaswami: p. 21 (top left)
- Alf Ribeiro: p. 11
- casa.da.photo: p. 9 (top)
- Chris Redan: p. 24 (top right)
- evgenii mitroshin: p. 21 (top right)
- Ilias Kouroudis: p. 9 (bottom)
- Jeppe Gustafsson: p. 22 (top left)
- Parikh Mahendra N: p. 8 (bottom right)
- Wirestock Creators: p. 8 (center left)
- YusufAslan: p. 18

All other images from Shutterstock

Crabtree Publishing

crabtreebooks.com   800-387-7650

In Canada: We acknowledge the financial support of the Government of Canada through the Canada Book Fund for our publishing activities.

Hardcover 978-1-0398-0659-7
Paperback 978-1-0398-0685-6
Ebook (pdf) 978-1-0398-0709-9
Epub 978-1-0398-0736-5

Published in Canada
Crabtree Publishing
616 Welland Avenue
St. Catharines, Ontario
L2M 5V6

Published in the United States
Crabtree Publishing
347 Fifth Avenue
Suite 1402-145
New York, New York, 10016

Library and Archives Canada Cataloguing in Publication
Available at Library and Archives Canada

Library of Congress Cataloging-in-Publication Data
Available at the Library of Congress

Printed in the U.S.A./012023/CG20220815

Contents

Do you wear cotton clothing? Cotton is a material used in many things. We use these things every day. But what is cotton? Where does it come from?

What Is Cotton?

Cotton is a soft, fluffy **fiber** found in nature. It grows on plants in many countries around the world. People use cotton fibers to make fabric.

Most cotton fabric is used to make clothing. Cotton clothing is strong and comfortable to wear. People also use cotton to make towels, blankets, curtains, and many other products.

COTTON FACT

About half of all fabrics are made from cotton.

Growing Cotton

Farmers plant cotton seeds in fields. The seeds grow into seedlings and then shrubs. White flowers bloom on the shrubs briefly. Then they fall off, leaving behind green pods called **bolls**.

COTTON FACT

Cotton grows in places with warm summers and mild winters.

Inside each boll, there are 20 to 40 cotton seeds and thousands of fibers. The fibers grow long and thick. About five months after planting, the bolls split open and fluffy cotton bursts out.

Harvesting Cotton

Farmers harvest cotton when it is ripe and ready to pick. In some places, workers harvest cotton by hand. They use their fingers to pull and twist the cotton out of the bolls.

WONDER WORD:
HARVEST

To harvest means to pick and collect a crop when it is fully grown.

Other farmers use big machines to pull cotton quickly from plants. Then they pack the cotton into tight **bales** so it can be stored or shipped.

Cleaning Cotton

Raw cotton contains seeds and bits of stems, leaves, and dirt. It must be cleaned well before it can be made into cloth.

WONDER WORDS:
RAW COTTON

Raw cotton is cotton that has been harvested but not cleaned or treated in any way.

Workers feed raw cotton into large machines. The machines break up and dry the cotton, then pull the fibers away from the seeds and other bits. Then they press the cotton back into bales.

Cotton Mills

Clean cotton is shipped to **mills**. Machines at the mills comb and line up the fibers. Then they stretch and twist the fibers, **spinning** them into yarn.

Next, machines called looms weave the yarn. They crisscross strands of yarn back and forth, joining them together to form solid pieces of cloth.

COTTON FACT

People add substances to yarn and cloth to make them different colors.

Cloth to Clothing

Large rolls of cotton fabric are shipped to factories. Workers check, sort, and store the cloth. Then people do different jobs to turn the cloth into clothing and other goods.

COTTON FACT

One bale of cotton can make 215 pairs of jeans, 249 bedsheets, or 690 bath towels.

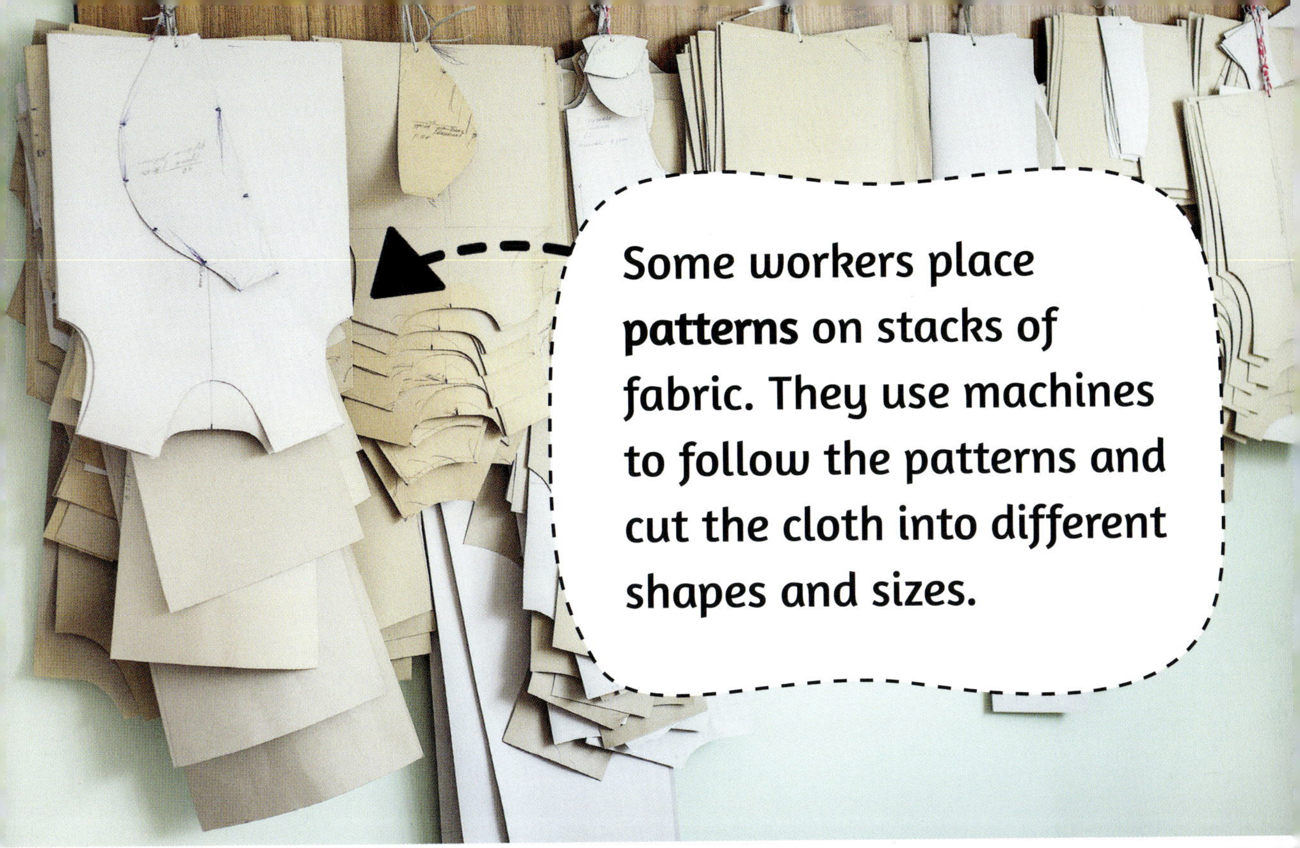

Some workers place **patterns** on stacks of fabric. They use machines to follow the patterns and cut the cloth into different shapes and sizes.

Workers use sewing machines to stitch pieces of cloth together. Each person usually sews one part of a pattern again and again to keep the work moving quickly.

Cotton Concerns

People around the world make their living by growing cotton and creating cotton goods we use each day. However, producing cotton can harm the environment.

Many cotton farmers use **pesticides** to kill beetles and other insects. The pesticides go into the air, soil, and water. Over time, they can make people and animals sick.

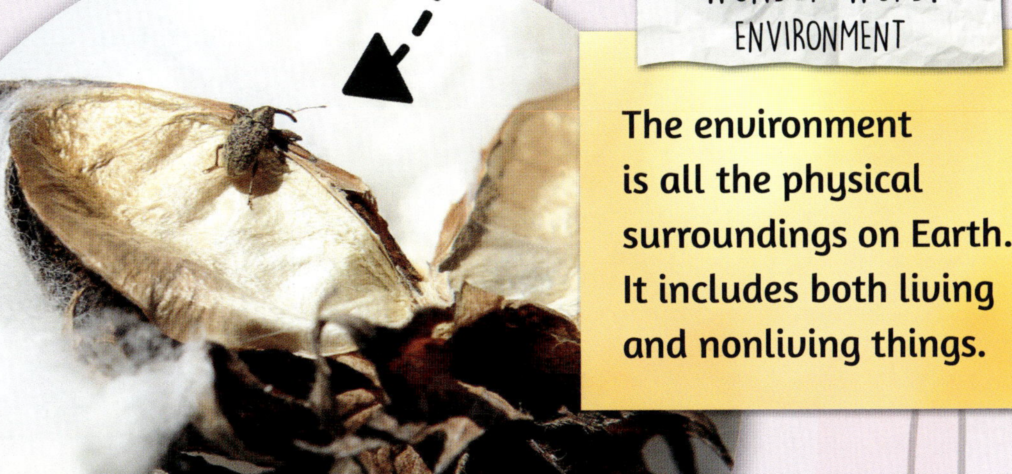

WONDER WORD:
ENVIRONMENT

The environment is all the physical surroundings on Earth. It includes both living and nonliving things.

It takes a lot of water to grow cotton. Many farmers do not have enough water for their crops, so they bring water from rivers and lakes to the farms.

When farmers use water for cotton, people may not have enough water for drinking, cooking, and washing. Bodies of water may shrink and even dry up.

COTTON FACT

The Aral Sea in Asia was a huge lake until people used the water to grow cotton. Now the lake is almost gone.

How Can You Help?

You can help by buying clothes and other items made from **organic** cotton. Farmers who grow organic cotton do not use harmful pesticides on their crops.

You can also help by buying used clothes instead of new ones. And you can give your old clothing, towels, sheets, and other cotton goods to people who need them.

COTTON FACT

You can save **energy** by hanging cotton clothing to dry instead of using a clothes dryer.

GLOSSARY

bale A bundle of cotton or other material that is tightly wrapped and tied

boll A rounded pod or capsule that grows on some plants

energy The power needed to do things such as work machines

fiber A thin substance that looks like thread or hair

mill A factory where cotton or other raw materials are made into usable products

organic Grown or produced without using human-made chemicals such as pesticides

pattern A model or guide for making something

pesticide A substance people spray on crops to kill insects that eat the crops

spinning The process of changing fibers into yarn or thread

INDEX